HOW PLANTS GROW

Angela Royston

Heinemann
LIBRARY

First published in Great Britain by Heinemann Library
Halley Court, Jordan Hill, Oxford OX2 8EJ
a division of Reed Educational and Professional Publishing Ltd.

Heinemann is a registered trademark of Reed Educational & Professional Publishing Limited.

OXFORD MELBOURNE AUCKLAND
JOHANNESBURG BLANTYRE GABORONE
IBADAN PORTSMOUTH NH CHICAGO

Designed by AMR Ltd.
Printed and bound in Hong Kong/China by South China Printing Co. Ltd.

03 02
10 9 8 7 6 5 4 3

ISBN 0 431 00200 2

British Library Cataloguing in Publication Data

Royston, Angela
  How plants grow. – (Plants)
  1.Growth (Plants) – Juvenile literature
  I.Title
  571.8′2

  ISBN 0 431 00200 2

Acknowledgements
The Publishers would like to thank the following for permission to reproduce photographs:
Ardea: p5, D Greenslade p22, J Mason pp6, 23, W Weisser p7; Bruce Coleman Limited: A Potts p4;
Liz Eddison: p26; Garden and Wildlife Matters: ppl2, 13, 14, 15, 16, 17, 19, 20, 21, 25, 27, K Gibson
pp9, 18; Chris Honeywell: pp28, 29; NHPA: S Krasemann ppl0, 11; Oxford Scientific Films:
K Sandved p8; Tony Stone Images: P D'Angelo p24.

Cover photograph: Ken Gibson, Garden and Wildlife Matters

The Publishers would like to thank Dr John Feltwell of Garden Matters for his
comments in the preparation of this book.

Every effort has been made to contact copyright holders of any material reproduced in this book.
Any omissions will be rectified in subsequent printings if notice is given to the Publisher.

Any words appearing in bold, **like this**, are explained in the Glossary.

Contents

Many kinds of plants

Plants keep growing until they die.
Some plants live for less than a year but
some trees live for hundreds and even
thousands of years.

All plants, even the tallest trees, start life as a tiny **seed** or **spore**. **Roots**, **stems** and leaves all grow from this tiny beginning.

Flowers and fruit

Many plants use **flowers** to make
new **seeds**. **Pollen** from these orange
flowers is carried by insects to **ovules**
inside other orange flowers.

The pollen joins the ovules to make new seeds. As the orange seeds swell and ripen, they are protected inside a juicy **fruit**.

Spores and cones

Some plants do not make **seeds** inside **flowers**. Ferns and mosses produce millions of tiny **spores** which are blown by the wind to start a new plant.

Conifer trees have **cones** instead of flowers. The seeds develop inside these woody cones.

A new plant begins

Inside each **seed** is a tiny plant which begins to grow when the seed is planted or falls into the soil.

First **roots** begin to grow down
and then a shoot pushes up
through the soil.

Roots

Green plants need water and sunlight to
grow well. **Roots, stems** and leaves all
play a part in keeping the plant alive.

Tiny hairs on the roots take in water and **nutrients** from the soil. Some plants have one big root, others have a tangled mass of roots.

Stems

The **stem** holds up the leaves and the **flowers**. These sunflowers have straight stems that grow long and tall to lift the leaves up to the light.

A stem has many tiny tubes which carry water from the **roots** to the leaves. This cactus plant also stores water in its fat stem.

Tree trunks

Tree trunks are hard, woody **stems**.
Trees need strong stems because they
grow much bigger and taller than
most other plants.

New wood grows every year so the trunk gets thicker and stronger. Bark is hard, dead wood which protects the growing wood underneath.

Climbing stems

Some plants have long, bendy **stems**.
These stems do not support the plant
but climb up something solid, such as
a tree or wall, instead.

This climbing plant has curly **tendrils** which twist themselves around a wire or the stem of another plant.

Leaves

Leaves use the energy of sunlight to make food for the plant from air and water. The plant turns its leaves to take in as much sunlight as possible.

The red tubes in this leaf bring water
from the soil and take away the
sugary food to the rest of the plant.

Evergreen leaves

Some trees are green all year round.
Holly trees have thick shiny leaves
that last a long time.

Conifer trees have small, pointed leaves like needles. These trees lose their leaves a few at a time.

Falling leaves

Most broad-leafed trees lose all their
leaves in autumn. The leaves may turn
yellow, red or brown before they fall to
the ground.

The tree rests during the cold winter weather. It starts to grow again in spring when new leaves unfold on the bare branches.

Storing food

Some plants store food in a **bulb** or swollen **root** in the ground. The leaves die back, but in spring they start to grow again.

The plant uses the food stored in the bulb until its new leaves begin to make food. This amaryllis needs lots of food to produce such a huge **flower**!

Which grows best?

Find out what plants need to make them grow. Put some damp kitchen roll in the bottom of three soup plates. Lay some mung beans on top.

Put the plates near a window. Let one plate dry out, add water to the second plate. Put the third plate in a box with the lid shut. After a few days, compare the plants. Which have grown best?

Plant map

a strawberry plant

leaf

flower

fruit

stem

roots

an oak tree

bark

leaves

roots

trunk

30

Glossary

bulb	swollen root which contains a store of food. Plants which grow from bulbs die back after flowering but grow again the following year.
cone	part of a conifer tree which makes new seeds
conifer tree	a tree which produces new seeds inside cones
flower	the part of a plant which makes new seeds
fruit	the part of a plant that holds the ripening seeds
nutrients	special things a plant needs to grow well
ovule	a female seed or egg cell. An ovule must be joined by a grain of pollen to become a fertilized seed.
pollen	grains containing male cells which are needed to make new seeds
roots	parts of a plant which take in water, usually from the soil
seed	contains a tiny plant before it begins to grow and a store of food
spore	the cell from which a new fern, moss or fungus begins to grow
stem	the part of a plant from which the leaves and flowers grow
tendrils	thin offshoots of the stem of a climbing plant which help to support the plant

Index